W9-CJM-201

the garden

EASY HEIRLOOM SEEDS

seed saving

FOR THE HOME GARDENER

guide

THIRD EDITION

Jill Henderson

Riverhead Free Library
330 Court Street
Riverhead, New York 11901

Groundswell Books
SUMMERTOWN, TENNESSEE

Library of Congress Cataloging-in-Publication Data is available upon request.

No part of this book may be reproduced or transmitted in any form or by any means whatsoever including photocopying, scanning, digitizing, recording, or any form of information storage-and-retrieval system, without written permission from the author or her heirs with the exception of brief quotations in articles or reviews.

Disclaimer of liabilities: Reference in the publication to a trademark, proprietary product, or company name is intended for explicit description only and does not imply approval or recommendation to the exclusion of others that may be suitable.

While every effort has been made to ensure the accuracy and effectiveness of the information in this book the author makes no guarantee, express or implied, as to the procedures contained herein. Neither the author nor the publisher will be liable for direct, indirect, incidental, or consequential damages in connection with or arising from the furnishing, performance, or use of this book.

© 2017 Jill Henderson

All rights reserved. No portion of this book may be reproduced by any means whatsoever, except for brief quotations in reviews, without written permission from the publisher.

Cover and interior design: John Wincek
Stock photography: 123 RF

Printed in the United States

GroundSwell Books
a div. of Book Publishing Company
PO Box 99
Summertown, TN 38483
888-260-8458
bookpubco.com

ISBN: 978-1-57067-346-7

22 21 20 19 18 17   1 2 3 4 5 6 7 8 9

MIX
Paper from
responsible sources
FSC
www.fsc.org
FSC® C005010

We chose to print this title on responsibly harvested paper stock certified by the Forest Stewardship Council,® an independent auditor of responsible forestry practices. For more information, visit https://ic.fsc.org/en.

contents

6 The Plant Profiles 34

Conclusion 54

the garden

EASY HEIRLOOM SEEDS

seed saving

FOR THE HOME GARDENER

guide

THIRD EDITION

introduction

Why Save Seed?

Thomas Jefferson said "The greatest service which can be rendered any country is to add a useful plant to its culture." If that's true, then the second greatest service would be to save the seeds of those useful plants.

But, besides being of service to ourselves and our country, why should the average person save their own seed? The easiest answer to that is, to save money.

I don't need to tell you how expensive garden seeds are, especially if they are organic. You could pay a small fortune to buy the seeds you need to grow even a modest size garden. With seed packs costing an average of $2.50 each, growing a garden large enough to sustain a family of four could cost hundreds of dollars a year. When you multiply the amount of money spent on seeds and plants by the number of years you plan to garden, the numbers grow even larger. By saving your own seed, you not only reduce the cost of growing your own food for that year, but for many, many years to come.

If you have never saved your own seed I think you might be surprised by how much seed can be reaped from just one type of crop. Have you ever eaten a watermelon? Can you guess how many seeds are inside of just one medium-sized fruit? That's a lot of seed to tuck away for your arsenal of self-sufficient food production, and all in a single season. Learning to save seed is a wonderful hobby and very rewarding, but it's also about survival. The seed you save from three watermelons would be enough to last you, your family and several friends and neighbors at least two years. Now, that's a lot of money saved!

But, while saving money is a perfectly good reason to save seed, there are several others worth mentioning.

By saving our own seed, we can preserve or enhance the characteristics of a particularly useful or interesting plant through the process of selection. Theoretically, seeds from plants that have grown successfully in a particular environment, such as your garden, are stamped with certain genetic markers that allow it to acclimatize to that particular circumstance. In other words, if you grow a squash successfully in your garden, the next generation grown from those seeds should grow as well or better than the parent plant in the same area.

Of course, I can't talk about saving seeds without discussing genetically modified organisms or GMOs—or, as I like to call them: Frankenfood. These Frankenfoods have had their natural genetic structure altered by mixing them with the genes of unrelated species—including those belonging to unrelated plants, animals, insects and even human genes. Now, I don't know what your spiritual beliefs are, but for me, I don't think God intended rice and mice to splice. And spirituality aside, there are serious concerns as to how these genetically modified plants act upon the human body when consumed. There are also very real concerns that these genetically modified organisms will pass their Franken-genes on to other, non-Frankenfood crops.

As a frightening example of this right here in the U.S., Monsanto has quite literally sued the land out from under farmers who were found to have even minute amounts of patented Roundup corn growing in their fields among their other corn. Monsanto sued because they said that these farmers were intentionally growing their patented corn without having paid for the right to do so. The farmers protested, saying that Monsanto's Round Up corn must be drifting onto unsuspecting farms from passing grain trucks, or from the seed mill where stray kernels were accidentally being mixed in with other corn seeds. But what they didn't realize was that these new Franken-crops were spreading their Franken-genes through the pollination process to non-GMO crops.

In one devastating example in mid-1998, University of Chicago scientists were experimenting with genetically modifying a variety of mustard to be herbicide resistant. And, although no known gene effecting floral characteristics was altered, the workers noticed that the genetically modified flowers looked a little different than those on the unaltered plants. Though the scientists thought this change was unlikely to be significant, they decided to test the modified plants' outcrossing rates, or the rate at which pollen successfully pollinates a female flower to produce viable seed in comparison to the non-altered plants.

It turned out that the genetically engineered mustard had over 20 times the outcrossing rate of the standard mustard. In short, the pollen from the genetically engineered mustard was over 20 times more likely to successfully reproduce than its natural counterpart growing right next to it! This disturbing fact spells disaster for non-GMO crops grown in the same region as open pollinated and organic crops, and its outcome has already been felt by farmers all over the Midwest.

According to the *Farmer's Guide to GMO's* written by David R. Moeller, in 2003 "GMO's were used by American farmers for at least 81 percent of their soybeans, 40 percent of field corn, and 73 percent of upland cotton."[1] That was five years ago! Today the numbers are even higher and include not just these crops, but many new ones, as well. It also includes consumer and animal products that have been made from or with them.

Even if you have a moral opposition to GMOs, you are probably eating them right now! Only certified organic foods do not contain genetically modified organisms.

Another issue related to GMOs is the patenting of life forms by the grain giants and the pharmaceutical industry. Make no mistake—the money to be made on the ownership of genetic patents is staggering. That's why the big grain and pharma giants like Cargill and Monsanto are racing to patent plant genes—and not just the GMOs that they created, but all plant genes of any value—like the vegetable crops that you and I grow in our gardens.

Imagine this scenario: all the plants in your garden are literally owned by a few multinational corporations. You must not only buy their seed, but you may have to pay for the right to grow it. And if you save seed or propagate the plants vegetatively, you

1. Farmers' Legal Action Group, Inc. (FLAG) and Michael Sligh—Rural Advancement Foundation International—USA (RAFI-USA), November 2004. www.rafiusa.org/pubs/Farmers_Guide_to_GMOs.pdf

may find yourself face to face with a lawsuit and a very real threat of going to jail for patent infringement. I'm not just talking about seeds from plants that these companies worked years to develop—I'm talking about old varieties that have been around for 50, 100, or even thousands of years! You can just see the dollar signs swirling, can't you?

The very real threat here is that, in the near future, we may not have the right to save our own seeds because some big corporate giant will own the rights to them. This frightening scenario is taking place all over the world this very minute.

In India, an American corporation claimed ownership to the indigenous neem tree through the patent process. Neem has been used in India for as long as history has been recorded for use as a food and medicine, and the leaves used as a natural insecticide. Despite that, this particular corporation sought to stop all people from producing the natural insecticide for their own use because they now owned a patent to it. Lucky for Indians, Vandana Shiva, a champion for working class people all over the world stood up to the corporate strong-arm tactics used to intimidate and subdue any resistance to their plans. Shiva helped the Indian people stand up against corporate bio-piracy, claiming indigenous rights to the use of the tree.

After a long and arduous court battle, the people won their suite against the American company, but only in India. Here in the US, this company still owns the patent and is the sole producer of neem oil. This scenario is played out over and over again all over the world. Sometimes the farmers win, but more and more often, they lose.

The big question is, when will this plant-grab come home to roost and how do we protect ourselves right now?

First of all, by saving and growing our own seed, we are essentially asserting our cultural and indigenous rights of collective ownership. The farmers in India won their suit in part because they could show that they had already claimed that particular genetic sequence simply by growing and using the plant for generations upon generations.

Secondly, by saving seeds of open pollinated and heirloom varieties of vegetable crops, we help ensure a naturally diverse gene pool within the various varieties of plants we choose to grow. The potato famine in Ireland was caused by a late blight that affected potatoes, which were the main vegetable food crop and the main source of income for many small farmers in the country. No potatoes in Ireland were immune from the blight because everyone was growing the same exact variety of potato. Had there been a larger variety of potatoes an immune variety could have been located and grown instead. As it was, hundreds of thousands of people suffered terribly and many more died of starvation for lack of a diverse gene pool.

That's all pretty heavy stuff, but unfortunately it is something we must consider when discussing the reasons why we should save seed. Whether you have a big garden, a little garden, a market garden, a farm, or a ten-acre food plot for self-sufficiency, knowing how to save your own seeds will save you tons of money—and one day, it may just save your life.

Seed saving is a fun and rewarding experience. It may seem a bit complicated at first, but with a little desire and a few basic instructions, saving your own garden seed is both easy and immensely rewarding. For those just beginning to save seed, there are numerous crops that can be easily saved with little additional effort in the garden. For those who have already dabbled in seed saving, my hope is that this guide will reveal some new and interesting information with which you can raise your seed saving skills to a higher level. By saving even a few of your own seeds, you not only help yourself, but you also help people around the world by preserving the genetic diversity of one of the world's most precious resources.

Happy gardening!

Jill Henderson

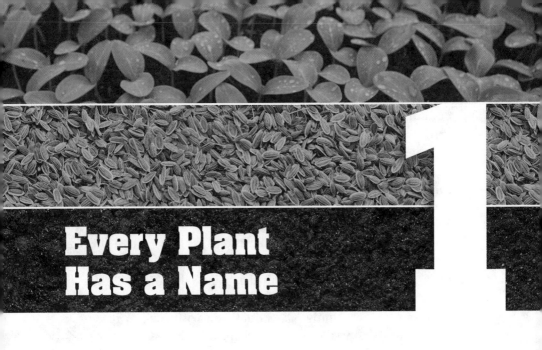

Every Plant Has a Name

knowing the Latin botanical names for plants is a very important aspect of seed saving. This type of classification is called botanical nomenclature. For some people, botanical nomenclature is one of the most fascinating aspects of plant propagation. For others, it is a thing to be avoided because it is full of long, hard to pronounce names. Too many people get hung up on the perfect pronunciation of these names and unless you are a botanist or have a serious reason to be able to say each name perfectly, I should think that pronunciation is trivial at best. The most important thing is that you know the name, not how to pronounce it. Knowing the name of the plants in your garden is important because they describe potential familial relationships between sometimes seemingly different plants and can immediately convey information on how a plant will grow and reproduce; both of which are important to the seed saver.

Family, Genus, Species, Variety

Just like us, all plants belong to a family. Their names, much like ours, can span many generations and include ancestral relationships. Plants are classified in descending order from kingdom, division, class, order, family, genus, species and variety. For our purposes, we need only focus on the family, genus, species and variety names. Take California Wonder bell pepper as an example. Its formal Latin name is written like this: (*Capsicum annuum var. California Wonder*), but may also be shortened to simply (*Capsicum annuum*).

FAMILY. A family is a large group of plants that all share certain characteristics such as the presentation or placement of flowers, leaves, stems, fruits, etc. What you don't see in the above example is the family name of this particular plant, which is often lacking on plant and seed labels. It is up to us to recognize which plant belongs to which family. Over time, you will come to recognize family members easily, but until then, use printed or online references as a teaching tool. I happen to know that peppers belong to the Solanaceae, or nightshade family, because I have learned to recognize the shape and presentation of its flowers. All nightshades, including tomatoes, eggplant, peppers and potatoes, share the same flowering characteristics. Through this simple process we are also able to see the relationships between plants that at first glance seem unrelated. One of the characteristics of the Solanaceae family is that most members have perfect, self-pollinating flowers. This information is enormously helpful when attempting to save its seed.

GENUS. Within each family are one to many genera (plural). Each genus (singular) represents a class of plants that share similar characteristics within the family. In our example, *Capsicum* is but one genus among several genera to which almost all peppers belong. The plants within the genus *Capsicum* produce fruit that are distinctly different in form than the other members of the Solanaceae family. Peppers belong to the genus *Capsicum*, while tomatoes and eggplant belong to the genus *Solanum* (formerly *Lycopersicon*). However, all of these genera still belong to the *Solanaceae* family.

SPECIES. Within each genus, plants are further divided by their similarities in structure or form. The species name is often a direct reference to an outstanding characteristic of that particular plant, the name of the person who bred or discovered it, or other such information. In our example, California Wonder bell peppers belong to the species *annuum*. It will soon become clear that *Capsicum annuum* includes almost every type of sweet or hot pepper known to mankind. And despite their different species names, even Kellu Uchu peppers (*C. baccatum*), Tabasco peppers (*C. frutescens*), and Chinese lantern peppers (*C. chinense*) will cross with *C. annum*. This fact tells us that almost all peppers are merely natural variations or cultivated varieties of the same species (in this case, the species *annuum*) and therefore can cross-pollinate pollinate with one another.

VARIETY. In our example, "California Wonder" is the variety of *Capsicum annuum* being discussed. Within *Capsicum annuum*, there are hundreds, if not thousands, of varieties of pepper. Giving each of these varieties a distinctive name helps distinguish it from other varieties or cultivars (short for cultivated variety). Ultimately, every variety within the *Capsicum annuum* group can cross-pollinate, or breed, with every other variety. Knowing this allows the seed saver to take steps to isolate different, but closely related, varieties of peppers from one another to prevent altering the characteristics of that variety by accidental crossbreeding.

2

Reproduction Essentials

perhaps it seems elementary to revisit a subject that most gardeners feel certain they understand; after all, biology is a subject we began learning about as children. However, even seasoned gardeners should take a moment to review the following information on plant reproduction—not because it is the absolute cornerstone of seed saving, which it is, but because it is fascinating, informative and fun.

Annuals, Perennials and Biennials

each of the plants we grow in our vegetable gardens is one of three types: annuals, perennials, or biennials.

ANNUALS are plants that move through their entire life cycle in one season: that is they leaf, set fruit and seed, and then die. Annuals must be planted every year from seed. Annuals include crops such as corn, lettuce, spinach, radish, legumes, melons and squash.

PERENNIALS grow year after year from roots that naturally overwinter in the ground. Perennial vegetables often begin producing fruit and seed the first year after being planted, but some may require two years or more before they will flower and set fruit. Some perennials, such as asparagus and fennel, are quite hardy and can survive very cold winters, but a few natural perennials grown outside of their native habitats (such as peppers and lemongrass) may not be able to survive a cold winter without human intervention. Despite being true perennials, these types of plants are most often grown as annuals and are referred to as *tender perennials.*

BIENNIALS grow from the same roots for two years and then die. Most tend to produce only leafy growth during the first season, which may or may not die back to the ground in winter. During the second season of growth, leaves again form, but this time the plant also produces flowering stalks, fruit and seed. Once it has completed its reproductive cycle, the plant dies a natural death and new seeds must be planted to start the two-year cycle all over again. Some biennials, like those in the Brassica family, have cold-sensitive roots that require special care to over-winter, while others, such as those found in the Apiaceae (formerly Umbelliferae) family, require little in the way of winter protection.

It is important to recognize that *most* garden crops are picked and eaten before they reach full maturity. In this sense, we treat almost all garden crops as annuals.

Take the cucumber for example. Immature cucumbers are picked while still small and green. Inside, the flesh is mild and crisp and the seeds are very small—exactly what you want in a table cucumber. But the seeds found inside an immature cucumber are not yet viable and will not germinate when planted. On the other hand, fully mature cucumbers are very large, yellowish-brown and often quite soft. The flesh of a botanically ripe cucumber can be spongy and bland, with big hard seeds that make the fruit unpalatable. But if you want to save viable cucumber seed, you will need to let at least a few fruits from several plants reach their botanical maturity on the vine. To learn more about when and how to pick fruits for seed production, refer to the "Harvest" section found in each plant profile later in the book.

Pollination

Before we can talk more about gathering fruits, we need to take a closer look at how those fruits are created in the first place. Pollination is the absolute crux of saving pure, genetically diverse seeds.

Without pollination, almost all garden plants will not produce fruit. A few exceptions include certain crops whose edible portions are produced underground, such as potatoes, onions, garlic and sweet potatoes. No pollination is required to obtain tubers, bulbs and cloves, which are not fruits, but a type of storage root. Pollination is also not required to obtain edible leafy greens, shoots or stalks of vegetative crops such as asparagus, lettuce, cabbage, broccoli, and so on. However, to obtain the seeds of these types of crops, pollination is absolutely essential.

Pollination occurs only when pollen from the anther of a male flower (or flower part) is transferred to the stigma of a female flower (or flower part). Once the stigma has been inoculated by a granule of pollen, a pollen tube begins to grow downward through the style to the ovary where one or more ovules are fertilized. Once fertilized, the ovary begins to swell and becomes the fruit. Inside the fruit, the fertilized ovules become seeds. When the seeds are mature, each will contain the germ (or embryo) of a new plant that will carry genetic traits inherited from both of the parent plants. Only fully ripe fruits bear mature seeds.

Pollination is typically achieved through the actions of wind, rain or insects, all of which move pollen from flower part to flower part. However, perfect, self-pollinating flowers have the incredible ability to pollinate themselves without the aid of outside forces, or even other flowers.

The term cross-pollination is defined as the movement of pollen between flowers, regardless of whether it is natural or mechanical. Natural cross-pollination (NCP) simply refers to the natural process of pollination, which only occurs between varieties of the same species. I will discuss cross-pollination in detail later on in this chapter.

How Plants Use Flowers

Although pollination is the physical act of reproduction, it is the arrangement of sexual organs within a flower that determines how reproduction occurs. There are many ways that flowers present their reproductive organs and knowing the basics of these mechanics is crucial to saving pure seed. For, while flowers are beautiful to look at, their sole purpose is to propagate the species.

All flowers are made up of several parts, the most noticeable being showy petals and sepals. The shape, color and scent of these flower parts are designed to attract specific pollinators like bees, butterflies and moths. Some flower petals even act as pollinator landing pads, directing insects deep into the heart of the flower where the sexual organs are located.

Stamens are the male sexual reproductive organs. Stamens consist of a thin filament that supports the anther, which produces pollen. Pistils are the female sexual organs. Each pistil is made up of a stigma, style and ovary. The ovary is essentially the womb of the plant where fertilized ovules grow into seeds. In the case of fruiting vegetable crops, the ovary is also the portion of the plant that becomes the vegetables we eat. For plants that don't produce recognizable fruits, but rather seed-bearing clusters, pods or kernels, the coating surrounding the seed is the fruit.

For the home gardener, it is important to know two things about flowers. The first is that each flower can be male, female or bisexual. The second is how pollination physically occurs. Some flowers need wind or insects to move pollen, while others require neither. Some flowers pollinate themselves before opening, while others will only accept pollen from flowers on a separate plant. Every seed saver understands that the seemingly elementary process of sexual reproduction in plants is never as simple as we might like it to be.

IMPERFECT FLOWERS (also known as incomplete flowers) are either all male or all female. These types of flowers need another flower of the opposite sex to complete pollination. Plants that have individual male or female flowers on separate plants are known as *dioecious* plants.

Like their flowers, dioecious plants are either all male or all female. Examples of dioecious plants include spinach and asparagus. In order to achieve pollination, dioecious plants must have at least one male plant of the same species growing near one or more female plants. Without a male, the females cannot produce fruit or seed.

Monoecious plants, such as squash, melons and cucumbers, have both male and female flowers on each plant. Male flowers are often the first to open and female flowers have tiny, but visible ovaries that resemble the fruit they will become.

PERFECT FLOWERS (also known as complete flowers) are bisexual—that is, each flower contains both male and female reproductive organs. Some perfect flowers can and do pollinate themselves. However, in some perfect flowers, the pistil will not accept pollen from a stamen within its own flower. These perfect flowers are known as *self-incompatible*.

Plants with perfect flowers include all members of the Legume and Nightshade families, such as tomatoes, eggplants, beans and peas.

Open-Pollinated, Hybrid and GMOs

Now that we have seen how different plants bear different types of flowers and how each of those flower types have their own requirements for completing sexual reproduction, we will turn our attention to the differences between open pollinated and hybrid plants.

OPEN-POLLINATED plants are those whose reproduction occurs naturally through wind, insects or other natural occurrences and not through human intervention. This term is most often used to describe plants that are the result of pollination between genetically similar parents of the same exact species and variety. Heirlooms and heritage plants are simply very old open-pollinated varieties. Naturally occurring hybrids can be the result of open-pollination (see cross-pollination below). However, for the seed saver, hybrids should never be thought of as open-pollinated plants.

CROSS-POLLINATION refers to the transfer of pollen from a male flower or flower part to a female flower or flower part. Nature only allows cross-pollination between varieties of the same species, whether intentional or not and natural cross-pollination (NCP) is a term used to describe that process. NCP occurs even when steps have been taken to isolate varieties within the same species.

Keep in mind that the effects of NCP, especially those represented by recessive genes, are not always obvious in the first or second generation of offspring. Many new vegetable varieties have come about because of unintentional cross-pollination, but for those interested in preserving a specific open pollinated or heirloom variety, cross-pollination between dissimilar varieties is undesirable. For plant breeders, intentional cross-pollination (or hybridization) is critical to developing new varieties.

HYBRIDS are the result of pollination between two or more genetically distinct parents (variety) of the same species. Hybridization can and does occur naturally. For example: if a *Homemade Pickle* cucumber is pollinated by a *Straight Eight* cucumber—both being varieties of *Cucumis sativus*—the resulting seeds and plants will all be hybrids.

Despite what anyone may say, it is entirely possible to save and grow seeds produced by hybrid plants. However, they will likely produce plants that do not resemble and will not produce the same type of fruit as the parent plant. In other words, they are unstable.

The only way to reproduce the characteristics of stabilized hybrids is by crossbreeding the same exact varieties together using the same exact process as the original hybrid parent. This is information that the breeder would rather not share. That being said, some plants that are labeled as hybrids are actually stabilized open pollinated varieties. When one of these fake hybrids is found, (this requires several years of growing-out to determine) its seeds will produce true to type offspring.

In the United States, all hybrid seeds must be labeled as such. This is often indicated by the word "hybrid" or "F1." The F1 indicates that the seeds in that packet are the first generation produced by intentional cross-pollination. The F stands for *filial* or *offspring*, so F1 refers to the first generation offspring after cross-pollination and an F2 is the second generation offspring, and so on.

Ultimately, open pollinated seeds will come true to form or variety only when pollinated by the same exact variety of the same exact species. For example: if a *Brandywine* tomato is pollinated by another *Brandywine* tomato, the seeds and resulting plants will be true to the *Brandywine* variety. If a *Brandywine* tomato is pollinated by a *Mortgage Lifter* tomato, the seeds and resulting plants will be a cross of the two parents, or an F1 hybrid.

For those interested in working on developing new or unusual characteristics in plants, or in creating new varieties through hybridization, I strongly recommend the intensely detailed and fascinating book, *Breed Your Own Vegetable Varieties* by Carol Deppe.

GENETICALLY MODIFIED ORGANISMS, or GMOs, can be any life form that has had genes of another life form, whether similar or not, artificially introduced (forced) into its gene sequence through the use of recombinant DNA techniques. Food crops that have been genetically modified are referred to as Genetically Modified Food (GMF). While most GMF plants produce viable seed that produce genetically identical plants, both forms are patented by the corporations that produced them. Anyone who saves the seeds of GMO crops are in violation of US patent laws and are aggressively routed out and prosecuted for patent infringement.

I am not an advocate of GMOs, believing the process to be both unnatural and harmful. I have yet to read a single scientific study not undertaken or sponsored by those with financial or political stakes in the matter that prove that food crops produced by genetic engineering are safe for long term use by either humans or animals. In fact, all of the studies done by truly independent researchers have been ignored, politically quashed, or publicly belittled as being "unscientific." Until we have honest, independent studies of the effects GMOs on humans, animals and the larger biotic community of naturally open pollinated plants, I cannot support or recommend them in any way, shape or form.

Preserving Genetic Purity

3

nless the aim of the gardener is to try to create new and interesting open-pollinated plants through the hybridization process, they will want to ensure that the seeds they save result in offspring exactly like the parent plants. This can only be achieved by maintaining control over the pollination process through the implementation of one or more methods of isolation.

Isolation is a term that refers to the separation of dissimilar varieties within a species through the use of physical or mechanical barriers to prevent undesirable cross-pollination.

Physical Isolation

ISOLATION DISTANCE refers to the physical separation of varieties that have the ability to cross-pollinate one another. Distances used to separate related varieties vary depending on the genus and species and whether the plants are self-pollinating or not. Commercial seed producers whose livelihoods depend on the purity of their seed tend to follow stricter isolation distances than are possible, or even necessary, for the home gardener. However, when attempting to preserve a rare heirloom variety to sell, or bank, research maximum isolation distance requirements and follow them as closely as possible. For the plants discussed in this guide, isolation distances vary from a few feet for bean varieties up to 800 feet for annual radishes. These distances are discussed in depth in each of the plant profiles later in the book.

BLOOM-TIME ISOLATION is the practice of planting same-species varieties at different times so that their blooming periods do not coincide. Individual sowings are timed in such as way as to allow the first variety to finish blooming before the second variety, planted later than the first, begins. This allows the seed saver to grow multiple crops of similar species without having to isolate the crops mechanically.

SINGLE-SPECIES ISOLATION is by far and away the simplest and most effective method of isolation. This technique requires the seed saver to grow only one variety of each species per season. This is particularly useful for plants that cross-pollinate readily, those that have large or showy flowers that are attractive to pollinators, such as okra, squash, melons and cucumbers, and plants that rely on airborne pollen, like corn and spinach. In this technique, the gardener will want to grow only one variety of any particular species during any given season. It is possible to grow multiple varieties of a single species that has perfect, self-pollinating flowers, such as beans, tomatoes and lettuce, provided that recommended isolation distances are followed.

Mechanical Isolation

CAGING is an isolation technique that excludes pollinating insects from transferring pollen from unwanted plants by enclosing the plants in a cage. Cage frames are often made of 1" x 1" lumber or plastic pipe and covered in fine breathable fabric, spun polyester, or wire mesh. Even tiny holes in the cage can allow the entry of pollinating insects and measures should be taken to seal all openings. In the event that the plants within the cage are not self-pollinating, pollinating insects must be introduced to the closed system. This latter technique can be both difficult and expensive for the average home gardener.

ALTERNATE DAY CAGING is used to allow several plants, or groups of plants, to be naturally pollinated by insects while preventing cross-pollination by caging one plant, or group of plants, on alternate days. Obviously, this allows the uncaged plant to be pollinated, while the other is isolated. This technique removes the need to introduce pollinators to an otherwise closed system and effectively allows natural pollination. The drawback to this method is that it is labor intensive and must be rigorously maintained during critical bloom times. Inadequate exposure to pollination can result in little or no seed production.

Either method of caging can be used throughout the entire bloom period and removed once all blooming has ceased and all blooms are producing fruit, or after a select number of tagged blooms are producing fruit and no further seed is desired. This is an excellent option for fruits that produce many seeds on one to several central flowering stems, such as lettuce, or on relatively compact plants, such as peppers. Caging does not work well on sprawling, vining crops like melons and squash—these plants are too large to cage effectively.

BAGGING is a form of caging in which individual flowers, flower heads, or clusters of flowers are covered with insect excluding materials like horticultural paper bags (for corn and spinach), spun polyester, pantyhose, muslin, or other obstructive, but breathable, material. Bagging is usually done in conjunction with hand-pollination and for plants whose pollen is wind blown. Breathable fabrics can be draped over entire plants or groups of plants, but they are most often formed into bags and used to cover one to several unopened flowers and secured with tape or string. Where multiple stems are gathered into a bag, cotton batting is stuffed around the stems at the bottom of the bag to prevent the entry of insects or the escape of pollen. Oftentimes, seeds or individual fruits are allowed to fully mature before the bags are removed. Among the crops commonly bagged are brassicas, spinach and squash. Spinach, like corn, has very fine wind-blown pollen that can travel up to 5 miles. This makes distance isolation for these crops almost impossible.

HAND-POLLINATION is the process of manually transferring pollen from a male flower part to a female flower part. For self-pollinating flowers, this involves making a dual-sexed flower into a single-sexed flower and then pollinating it by hand. This process is quite difficult and will not be covered here. For all other flowers, hand-pollination involves locating individual flowers or flowering stems that are just about to open, taping them closed and in the morning, transferring the pollen by hand.

Knowing how plants present their flowers beforehand is important, because locating the different sexes in unopened flowers can be difficult. Male flowers used in hand-pollination can be from the same plant as the female flowers (*selfing*), or from a different plant of the same variety (*sibbing*). Selfing is recommended only when the purity of the variety is in question, or when only one plant is available; otherwise, sibbing creates genetically stronger plants.

Evening is the best time to search for blossoms to hand-pollinate. Select only those that appear ready to open the following day. These buds are often noticeably swollen and show streaks or blotches of color near the base of the petals. For plants that have individual male and female flowers, such as squash, both sexes of flowers must remain unopened prior to hand-pollination to prevent contamination. Never use flowers that have begun to open, or those that have been chewed opened by aggressive pollinators, like bumblebees. Once the appropriate unopened flowers are located, they should be tagged by tying a bit of colored string to the stem before the tips of the flowers are taped closed with ¾" masking tape to prevent the flowers from opening prematurely. In sprawling crops such as cucumbers, melons and squash, tall stakes can be used to mark the location of flowers, making them and their fruits easier to find later on.

Work with one female flower at a time to avoid insects slipping in when you least expect them. Early the following morning, snip off the male flower(s), leaving a length of stem attached for a handle and bring them to the female flower(s). Tear away the petals of the male flower to reveal the stamens and hold them by the stem between your teeth while gently removing the tape from one female flower. Allow a few moments for the petals to reflex open and gently rub the anthers around the style of the female flower. Carefully re-tape the petals of the female flowers, being careful not to cover the sepals or the ovary. Otherwise, pollinated flowers can be bagged securely. If bags are used, allow the fruit to begin forming before removing them.

Natural Cross-pollination

I can't stress enough the importance of controlling the pollination process when dealing with same-species plantings. In relation to the concept of saving pure seed through various isolation techniques, plants can, do, and will naturally cross-pollinate. As seed savers, we would like to believe that we can control the microenvironment that precludes outside pollination, but nature has a way of getting one over on us through its own mechanisms and it occurs more often than most people realize.

When planting crops for seed, it is important to remember that NCP is enhanced by the presence of common pollinator insects such as bees, sweat bees and bumblebees, as well as weaker pollinators such as flies, wasps and other insects. Even with the perfect, self-pollinating flowers of tomatoes, NCP occurs at a rate of 2-5% when plants are grown closely together.[1] To help reduce the incidence of NCP in seed crops, it is best to plant each variety in large blocks, not long narrow rows, as is common in the home garden. This planting scheme increases the chances that the blossoms in the center of the block will be pollinated by its closest neighbors, i.e. the same variety.

Another way to avoid NCP is by planting rows of dissimilar species between blocks of seed crops. For example; growing a tall flowering crop, such as beans, in between two different varieties of tomato plants will help ensure that pollinators stick to one side or the other, reducing the chances of cross-pollination. It helps if the separating crop has flowers blooming at the same time as the seed crop and that these flowers be more attractive to active pollinators.

Despite their best efforts, all seed savers experience the effects of natural cross-pollination at one point or another. Even after isolation techniques are used and all care has been taken to avoid it, the very best defense against the loss of a pure seed variety is to always, always keep a supply of pure seed in storage as a backup.

1. McCormack, Jeff, Ph.D., *Guidelines for Maintaining Purity in Pepper Varieties.*

4

Selection and Genetic Diversity

Genetic Diversity

All of the prior information on pollination and isolation is intended to teach the seed saver about preserving the genetic purity of plant varieties. However, preserving genetic purity goes hand in hand with maintaining the proper levels of genetic diversity. After all, genes are responsible for every aspect of a plant's growth, production and reproduction. Genes determine a plant's physical characteristics, such as height, leaf shape and fruit size. They also play a role in resistance to disease, drought and insect pests, among other things.

Genetic diversity also makes it possible to have so many wonderful varieties of plants in the garden and it is genetic diversity that helps keep those plants vigorous and healthy. Each and every tiny seed contains an amazing array of genetic material that has been passed down from generation to generation.

Saving genetically pure seed is one way to insure that offspring grown from them will look and act just like the parent plants. However, to attain the same level of health and vigor as the parent plants, the gardener must focus on saving save seeds that also contain a deeply diverse pool of genes—even those that are recessive. Therefore, it makes perfect sense to talk about how one can maintain, or even improve the genetic diversity of the seeds we save.

Genetic diversity isn't something most gardeners talk about, but it is definitely one of the most important aspects of saving seed—and it's so easy to do! Through a process known as *selection*, it is possible to save seed that is genetically diverse.

In this chapter we will discuss the three aspects of maintaining genetic diversity: selection, roguing and inbreeding depression.

Selection

When an open-pollinated or heirloom variety is grown for the purpose of gathering seed, the seed saver would do well to focus on gathering seed from those plants that exhibit outstanding characteristics typical of the variety, while culling (rouging) out those that do not. This is the process of *selection*.

The characteristics of a plant are essentially its distinguishing traits or character. These traits encompass all of the physical aspects of growth—from the moment germination occurs to the final ripening of fruits—and includes such things such as germination time, ability to grow in hot or cold weather, overall plant size or form, fruit shape, flavor and maturation time, as well as tolerance or resistance to disease, drought and insects. These are but a few traits to base your selection on.

For instance, if you prefer your tomatoes to ripen all at once, then you should try to save seed from as many of those plants that did just that, while avoiding those that did not. If some of your lettuce plants didn't bolt at the first hint of hot weather, then you might want to consider saving seeds from those particular plants. By doing so, you increase the chances that the seeds you plant next year will contain the genes responsible for that late-bolting characteristic. However, saving seeds from just the plants you like is not enough.

Roguing

Merriam-Webster defines a rogue as "an individual exhibiting a chance and usually inferior biological variation," which is the perfect description of plants that do not conform to the "typical" characteristics of any given variety. These types of plants often have a disagreeable physical anomaly, such as a plant that is smaller than all the others or one that has misshapen leaves or fruits. That is not to say that every plant should be a clone of the other, rather, it is about those plants that exhibit differences that are both noticeable and undesirable.

For example: Your tomato seeds have just germinated. Of all of the new seedlings, one or two have deformed leaves. This could simply mean the leaves had a hard time sloughing off the seed coating. However, if the deformation persists into the first set of true leaves, these plants should be considered rogues—an inferior biological variation from the norm. Plants such as these should be culled from the group as quickly as possible to prevent their flowers from contributing unwanted traits to the collective gene pool (i.e. the seeds you save). This is called *roguing* or *roguing-out*.

With this in mind, it is not always necessary or desirable to save seed from only the "very best" plants, nor is it a good idea to save seed from only one or two plants. Rather, seed should be saved from as many plants that exhibit the positive characteristics of the variety as a whole. Remember, the goal is to increase genetic diversity, not limit it to a narrow list of preferences. If you want clones, take vegetative cuttings. If you want a vibrant and diverse group of plants, save seed from as many different plants as possible. Failure to follow this advice may result in a condition known as *inbreeding depression*.

Inbreeding Depression

Inbreeding depression is a fairly self explanatory term, but in short, plants that are grown from seed that has been saved from too few plants can result in offspring that are not as genetically diverse as their parents.

Inbreeding depression results in the loss of genetic diversity, which can in turn result in loss of vigor in germination, growth, fruit production and disease tolerance. However, not all plants are affected the same. Plants with perfect self-pollinating flowers, such as beans, peas and tomatoes, rarely suffer from inbreeding depression, while others such as corn, are highly susceptible.

For example, cucumbers are somewhat affected by inbreeding depression, so seed should be saved from at least 1 fruit from each of 6 individual plants. Corn, on the other hand, requires saving seed from a minimum of 100 out of a stand of 200. To find out about the specific requirements for avoiding inbreeding depression, look under the heading, *Pollination & Isolation* in the individual plant profiles.

For the home gardener, a good rule of thumb is to save seed from at least the best two thirds of any one variety in order to maintain a general standard of diversity, and up to one half for overall improvement of the variety.

During the selection process, those plants or fruits that have been chosen for seed production should be marked with colored string or other markers as early as possible in the growing season so that those fruits are not mistakenly picked for table use.

Of course, you don't need to actually keep all of the seeds you produce. After all, the amount of seed obtained from 6 cucumbers adds up very quickly. However, all of the seed collected should be mixed together before removing the portion of seed that you do not want to keep. This way, the seed saver is sure to have a good blend of genetic material with which to maintain a wide base of genetic diversity within the variety. I don't like to throw anything away, so I usually dry all of the seeds I collect and give away the extras.

To grow out varieties requiring high levels of isolation, I highly recommend working in tandem with other seed savers. If the collective group lives within the same region, the crops that are grown will all have experienced similar, but different, growing conditions that will be favorable to everyone involved.

If, for example, you would like to grow and save seed from acorn squash and zucchini—both varieties of *Cucurbita pepo*—it would be very difficult to do so without hand-pollinating. But if two people were to cooperate on the effort, one could grow the acorn squash while the other grows the zucchini. This way, both participants can have fresh produce to eat and pure seed to sow in the years to follow.

Working as a group to save a single variety also works when space is limited. For example; you would like to grow and save seed from a variety corn, but do not have the space for 200 corn plants, which is the minimum number of plants to grow to avoid inbreeding depression in that crop. If two gardeners each grow 100 plants and save seed from 50 plants each, the seeds from all 100 plants can be combined to maintain the genetic diversity of that variety. Each gardener will have all the seed they need for several years and still have plenty of corn for table use.

Besides being a good way to maintain pure seed strains, working with fellow seed savers can be a fun and rewarding way to develop new and lasting friendships.

Harvesting Seed

5

a s gardeners, the most rewarding part of the season is when the harvest begins. The same is true for seed savers. After a long season of planning and monitoring your garden crops, the reward is more than bountiful.

How to Determine Maturity

W hen saving seed, botanical maturity is the goal. Almost all of the vegetables that we put on our tables are immature versions of the botanically ripe fruit. For example, cucumbers, okra, lettuce, radishes and peas are all picked for consumption before the fruits have had the opportunity to develop or ripen their seeds. This phase of growth is referred to as market maturity. In addition, some crops, like biennial brassicas, may need two full years of growth before they will reach botanical maturity and produce seed. Certain plants, such as tomatoes, melons and beans produce copious amounts of fruit, which often allows the gardener to consume a portion of the crop and still have plenty of fruits left to produce seed.

On the other hand, produce from other crops like broccoli, radishes, turnips and head lettuce must be left completely intact if seed is desired. Very few of the vegetables we eat remain on the plant long enough to produce viable seed and those that do are often quite unappealing as food. For example, cucumbers are enormous, yellow and very soft when ripe. Specific details on maturity for each species can be found in the plant profiles.

Collecting and Cleaning Dry Seeds

Dry seeds are those produced in pods or on flowering stems and include all brassicas, lettuces, root crops and legumes. For seeds produced on flowering stems, the seeds are ripe when they begin to turn from green to brown or black, or when the small fruit capsules begin to open. These seeds should be gathered before natural dispersal, or shattering, occurs. This can be accomplished by either gathering the entire stem or cluster as the seeds ripen, or by simply shaking the ripe seeds into an open bucket or paper bag daily. Otherwise, the entire plant can be gathered, hung upside down inside a paper feed sack, drop cloth or bucket until the seeds are dry. For legumes, seeds are ripe when they can be heard rattling inside the pods. Because these types of seeds are already hard and almost fully dry when mature, cleaning them is as simple as removing the seeds, winnowing the chaff and storing.

CLEANING SCREENS are useful when processing large amounts of seed whose final appearance is important, such as when selling seeds. Frames are mounted with screens that have openings of various sizes. Seeds and chaff are placed on the top and sifted or pushed through the screens. In most cases the openings in the screen are used to allow the seeds to fall onto a waiting tarp or into a bowl, while the chaff remains in the screen.

THRESHING is the process of separating seeds from other plant materials such as stems, leaves and pod shells. This can be accomplished by flailing the dry coverings or seed heads to release the seeds and is often done over a container, on a large tarp or sheet, or inside of one. Flailing can be as physical as beating the material with a blunt object or stomping or crushing them with a hammer, or as gentle as rubbing seed heads between gloved hands.

WINNOWING is the process of separating heavier seeds from lighter chaff. In this process, the relatively heavy seeds fall straight down, while the lighter chaff and immature seeds blow away. This is often accomplished by pouring material that has been threshed from one vessel to another, allowing the contents to fall freely through lightly moving air. Other winnowing methods include tossing the contents into the air from a basket and catching the falling seeds (fun, but a bit tricky at first), by sifting the chaff through various sized screens, or by using a batten and tarp or board that has been set on a slight incline and allowing the heavier seeds to roll down to the bottom while the chaff is held back by the batten. For home gardeners and seed savers, removing all of the chaff is not necessary to the function or preservation of the seed itself.

Cleaning Wet Seeds and Fermentation

Wet seeds are all those that are borne within a moist fruit. Melons, tomatoes, squash, peppers, eggplant and cucumbers are all examples of wet seeds. These seeds must be removed from the moist flesh and membranes of the fruit, and cleaned and dried before being stored. Most wet seeds are simply rinsed in water to remove membranes, chaff and sticky fruit juices. Because fruit-borne seeds are at varying stages of maturity even when the fruits appear completely ripe, it is often a good idea to separate those that are viable from those that aren't. This is easily achieved by placing the cleaned seeds in a large bowl of water in which almost all of the mature seeds will sink rather quickly, while the immature ones float. Once the seeds have been sorted, drain them immediately in a mesh colander that has been placed on a dry towel or paper toweling. Dry the seeds further by spreading them out in a single layer on rigid plates or pans that have been covered with paper coffee filters to prevent sticking. Keep drying seeds out of direct sunlight until they are crisply dry.

While most seeds require simple rinsing, a few, like tomatoes and cucumbers benefit from a wet cleaning process known as fermentation. This process uses a fungus to help eliminate seed-borne viruses and to break down gelatinous coatings on the seeds, which if left intact, may inhibit germination.

In a jar, place the seeds and enough natural juice from the fruit to cover them. If necessary, a small amount of water may be added to the mixture, keeping in mind that water may actually stimulate some seeds to germinate. Stir the mixture vigorously and set away from direct sunlight in a warm area between 65°–75° F. for several days to one week.

After a day or two a fungus will form on top of the mixture, after this happens the contents should be stirred gently once a day to promote the breakdown of pulp, gelatinous coatings and to separate seeds. Once the fungus forms, the seeds should only be allowed to remain in the jar for two to three days, maximum. Individual seeds can be taken out and rinsed thoroughly in water to check the dissolution of the gelatinous covering. Once the gelatinous coating becomes thin and weak, fill the jar ¾ full of water, cap and shake briskly to help remove lingering coatings and to separate seeds. After shaking, allow the contents to settle for one minute. During this time the mature seeds will sink to the bottom, while pulp and immature seeds will float to the top. Carefully pour off the excess water along with the floating chaff and immature seeds, leaving the mature seeds in the jar. Again, fill the jar with water, shake and pour off chaff, repeating the process until the water is clear.

Once the seeds have been sorted, drain them immediately in a mesh colander that has been placed on a dry towel or paper toweling. Dry the seeds away from direct sunlight on rigid plates or pans covered with paper coffee filters to prevent sticking.

Drying and Storing Seeds

All seeds have a finite lifespan, but by properly cleaning, drying and storing them, the home gardener can preserve germination rates beyond the average for each variety. Seed banks employ rigid protocols to prolong a seed's viability by maintaining constant temperatures and specific humidity levels and most national seed repositories store many of their seeds at subzero temperatures.

While home-based seed savers may not be able to achieve as high a level as this, they can still store seeds for an incredible length of time if a few measures are taken to control environmental conditions.

The first step to ensuring the longevity of seeds is to dry them completely before storing. Seeds that are not dried and stored properly will not achieve true dormancy and will continue to respire, or breathe, exhausting energy that the seed needs to germinate and sustain early growth. Most seeds dried at room temperature will have a moisture content of 10%–20%. At these levels, seeds that are packaged in glass or plastic can rot, mold or germinate prematurely. Those that do survive storage tend to have reduced germination rates.

The only seeds that dry well enough to be frozen without special treatment are tiny, hard seeds like eggplant. Room dried seeds should only be wrapped in acid-free paper packets, which can then be grouped into a large rigid box and stored in a dark area that remains below 50° F. Stored this way, most seeds should maintain at least 50% germination rates for several years.

The ideal moisture content for most seeds in storage is between 6–8%, at which point seeds can safely be packaged in airtight jars or plastic before being stored or frozen. Seeds should never be dried in the oven, dehydrator, microwave, or in direct sun, as high temperatures can kill the embryo within the seed. The safest, easiest way to achieve satisfactory drying is through the use of silica. Once seeds have been room-dried, they are wrapped in paper packets and placed on top of an equal amount of silica that has been spread in the bottom of an airtight container. The seed packets are left in the container for 2–3 days before being removed.

After drying, seed packs may be enclosed in airtight jars or plastic bags and frozen or kept in the refrigerator. Never store seeds in or with silica gel. The silica tends to absorb too much moisture, killing the embryos within the seed.

If seeds are kept in the refrigerator or freezer, take special care to allow all containers to come to room temperature before opening. When a very cold container is brought into room temperature, condensation forms on the outside. If the container is opened when cold, the condensation will also form on the inside, potentially rehydrating the seeds within.

Storage Times for Seeds

nce your seeds have been harvested, dried and stored, it is important to remember that they are living things and their life spans are finite. In other words, they won't remain viable forever. In perfect storage conditions seeds will last anywhere from one to six years, depending on the type of seed. If storage conditions are average to poor, germination rates could possibly be reduced by as much as 50%—that means that up to one-half of all saved seeds could fail to germinate after being planted. That's a lot of seeds to lose if you are depending on a crop. The following table of storage times is a helpful guide for storage lengths of some common food crops. Some crops are listed in more than one group, either because of differences in species or cultivars, or in the natural variances in factors that determine seed viability.

SHORT (1–2 YEARS): Anise, caraway, chive, cumin, leek, lovage, marjoram, onion, oregano, parsnip, peanut, salsify, sweet cicely.

MODERATE (3–4 YEARS): Beans (*Phaseolus species*), broccoli, carrot, celery, celtuce, chervil, corn, eggplant, fennel, ground cherry, leek, lettuce, parsley, parsley root, peas, pepino (melon pear), peppers, potato, tomatillo, tomatoes.

INTERMEDIATE (4–5 YEARS): Asparagus, basil, dill and okra. And Brassicas such as, broccoli, broccoli raab, Brussels sprouts, cabbage, cauliflower, Chinese cabbage, Chinese mustard, collards, garden cress, kale, kohlrabi, leaf mustard, mustard greens, radish, rocket, turnip and rutabaga.

LONG (6+ YEARS): Artichoke, beans (*Vicia* and *Vigna species*), beets, cardoon, celeriac, celery, chicory, cucumber (common, burr and Indian gherkin), endive, gourds, melons, orach, quinoa, spinach, squash, sunflower, Swiss chard.

Testing Germination

any seed savers fail to test their seeds for germination before planting in the field. This is a vital mistake that is often paid for over and over throughout the course of the growing season. By testing germination rates before planting ever begins, the seed saver will not only know how heavily to sow a particular type of seed, but whether or not the seed retains enough genetic diversity to maintain healthy germination rates. Reduction in germination rates is among the first indications that a variety has lost a large portion of its genetic diversity. Should this be the case, more of these crops can be grown and more seeds from more plants can be saved the following year to try to improve the situation.

To test germination, wet two layers of thick paper toweling with water until moist, but not dripping. Place an even number of seeds on one of the towels (even numbers are easier to derive percentages from). Fifty seeds is good, but less will do.

Cover the seeds with the second towel and press out the air before sealing the seeds in a large zip-top bag. Seeds need to breathe, so leave one corner of the bag open and write on it the average number of days to germination for the species. Place the bag away from light in a warm area averaging 65°–70° F. Every two days count how many of the seeds have germinated and write the numbers down.

Since most seeds germinate within 10–14 days, pay special attention to how many seeds have germinated by 7th, 10th and 14th days. Any seed that germinates after 14 days should be considered to have poor germination and not counted.

Using the total number of seeds that have sprouted by the 14th day, we can determine germination rates expressed in percentages. If 25 out of 50 seeds germinated in that time, the germination rate would be 50%, which is not necessarily bad, but it's not good either. Technically, seeds with germination rates below 50% are considered to have very poor germination and work on the variety or seed storage conditions should be examined more closely. For the home seed saver, a very good germination rate would be in the high 80% range.

6 The Plant Profiles

LEVEL ONE easy

The seeds that are easiest seeds to save include all legumes, lettuces, peppers, eggplants, tomatillo, tomatoes, and annual radishes, most of which are self-pollinating and produce viable seed in one season. It is surprising how many different varieties can be grown in a single garden with little more than distance isolation practices. Pay attention to the genus and species names in italics to see which plants are likely to cross-pollinate.

ASTER

Asteraceae Family (formerly Compositae)

Lettuce *(Lactuca sativa).* Includes and will cross with all varieties of lettuce including loose leaf, butterhead/bibb, crisphead/iceberg, romaine/cos, celtuce/asparagus lettuce, and wild lettuce.

The six most common types of lettuce are:

Crisphead/Iceberg—head lettuce.

Butterhead/Bibb—small loose green heads and soft leaves.

Romaine/Cos—elongated leaves form upright loaf shaped head.

Loose Leaf—produces soft rosettes.

Celtuce/Asparagus lettuce—thick succulent stems and leaves.

Latin Lettuce—loose head with elongated, somewhat leathery leaves. Primarily represented in the U.S. by the variety "Fordhook."

POLLINATION & ISOLATION

The only annual members of the Asteraceae family in the vegetable garden are lettuce, celtuce, and sunflower. The remaining members, of which there are many, are either seed producing biennials or tuberous perennials. Therefore, only lettuce and celtuce are covered in this book.

Lettuce and celtuce have perfect self-pollinating composite flowers. Because their fused anthers shed pollen inward toward the stigma, lettuce rarely cross-pollinates. However, because the flowers open for thirty minutes to three hours every morning, there is a slight chance for cross-pollination by insects. Therefore, it is a good practice to separate varieties by 10–20 ft. or by a taller crop to ensure seed purity. Otherwise, the flowering stalks can be enclosed in large blossom bags or closely wrapped with lightweight spun polyester supported on stakes. Do not harvest leaves from heading type lettuces for table use at any time during growth. A light harvest of outer leaves from loose-leaf types may be taken without harming seed production.

GENERAL HARVEST

Because individual clusters of seed are pollinated at different rates, seeds also ripen at different rates. Collect ripe seed daily by shaking the clusters over a bowl or open paper bag. You can also pluck the seeds from the capsules, though this is more time-consuming. Otherwise, remove flowering stalks after most of the seedpods begin turning brown and place them upside down in a paper bag to finish drying. Rub seed pods between the hands to release seeds and screen or winnow to remove chaff.

OTHER

Lettuce is one of the few garden crops that does not display inbreeding depression, which means that it is entirely possible to save viable seed from just one plant. However, genetic diversity is strengthened by saving seed from at least 5–10 plants, and up to 20 plants is best for preservation of a rare variety. Rogue out weak or small plants and any that are off-color or shape for the variety. When red and green lettuces cross-pollinate, red dominates. This results in primarily red-leaved hybrids, which should be rouged out immediately. The flowering stalks of head lettuces emerge from within the head itself. Once the heads begin to soften, use a sharp, clean knife or box cutter to make a 2" deep X in the top of the head. This not only helps release the flowering stalk, but prevents head rot, too.

BRASSICA

Brassicaceae Family

Summer Radish *(Raphanus sativus)*. This level includes only common annual summer radishes and edible-podded radishes (formerly *R. caudatus*), which will cross with all varieties of summer, winter, and wild radishes *(R. raphanistrum)*. Winter radishes are biennial and are not covered in this guide. All radishes will cross-pollinate one another, but will not cross with other Brassica family members, such as mustard, cabbage, turnip and kale.

POLLINATION & ISOLATION

All radishes have perfect self-pollinating flowers that are self-incompatible, which means that pollen from the male flower organs will not fertilize the female organs on the same plant. Because insects must move pollen from one flower to another, cross-pollination between varieties of summer, winter and wild radishes will occur.

Isolate different radish varieties by at least 800 ft. for home seed savers and to up to a mile for preservation of rare varieties. Many different varieties of annual radishes may be grown during the season for table use, but allow only one variety per season to flower for seed production. Plant seeds using traditional spacing.

GENERAL HARVEST

At market maturity, dig up roots for examination, leaving the plants intact. Eat the roots that are not wanted for seed and cull those with roots that don't conform to the variety or are diseased. In order to accommodate their large size during flowering, replant quality roots to stand 4"–6" apart in the row with 24" between rows. The many-branched flowering stalks reach upwards of 4' tall and often need staking. Seeds may be red, brown, yellow or black, but save all colors for the best genetic diversity. Dry pods do not readily shatter. Pick individual pods or branches as they begin to brown and allow them to dry completely before threshing. Pods are hard and sharp, so wear gloves. Lay pods between layers of heavy tarp and stomp on them with heavy boots, or place pods in a large sturdy bucket and crush them with a blunt object like a 2x4. Winnow away the chaff and store.

OTHER

Once the main seed crop is re-established in the garden for seed production, roots should not be harvested for the table. Harvesting some of the early green, edible seedpods won't hurt the seed crop though. Avoid inbreeding depression by saving seed from at least 6 plants for the casual seed saver and 20–50 for long-term preservation of the variety.

LEGUME

Fabaceae Family (formerly Leguminosae)

Garden Bean *(Phaseolus vulgaris)* will cross with all varieties of bush, half-runner, pole, green, snap, shelling and dry beans.

Lima Bean *(Phaseolus lunatus)* will cross with all varieties of lima beans.

Runner Bean *(Phaseolus coccineus)* will cross with all varieties of runner beans, including white-seeded varieties incorrectly referred to as "limas."

Tepary Bean *(Phaseolus acutifolius)* will cross with all varieties of tepary beans.

Fava Bean *(Vicia faba)* will cross with all varieties of fava, broad, bell and horse beans.

Cowpea *(Vigna unguiculata)* will cross with all varieties of cowpeas, crowder beans and yard-long beans.

Garden Pea *(Pisum sativum)* will cross with all varieties of garden, snap, edible pod, snow, shelling, field and dry peas.

Soy/Edamame *(Glycine max)* will cross with all varieties of soy and edamame.

Lentil *(Lens culinaris)* will cross with all varieties of lentils.

Chickpea *(Cicer arietinum)* will cross with all varieties of chickpea.

POLLINATION & ISOLATION

Legumes have perfect self-pollinating flowers, but will often cross-pollinate in the presence of heavy pollinator activity and on large-flowered varieties. For home gardeners, separate varieties by planting in blocks or rows that are 10–20 ft. apart. For rare heirlooms, isolate by at least 50 ft., preferably with a taller or flowering crop in between.

GENERAL HARVEST

Beans reach botanical maturity approximately 6 weeks after market maturity, while peas take about 4 weeks. All legume seeds are ready to harvest when they can be felt or heard moving inside the pod. If most of the pods have turned brown and threat of frost, excessive moisture, or predation at harvest time are a concern, the plants can be pulled up by the roots and hung in a cool, dry place until pods dry completely. Once dry, pods can be flailed or shelled by hand and winnowed to remove chaff. Continue to dry seeds out of direct sunlight until they shatter when struck with a hammer. If weevils are a problem, freeze fully dried seed in a plastic container for five or six days before packed for long-term storage.

OTHER

Rogue out all plants that show disease, poor growth, and unusual characteristics. Viable seeds can be saved from just one plant, but for long-term preservation and avoidance of inbreeding depression, saving seed from 5–10 plants is best.

Sow seed as you would for table use, but mark plants meant for seed and do not harvest any pods for eating until you have harvested all you want for seed. Do not wait until the plant is almost done producing to allow pods to set seed or you will be selecting for late fruit set. Desirable characteristics for legumes might include size and number of seeds, vigor, upright habit, disease, drought, and insect resistance.

MALLOW

Malvaceae Family

Okra *(Abelmoschus esculentus)* will cross with all varieties of okra.

POLLINATION & ISOLATION

Okra has large, perfect self-pollinating flowers that can be cross-pollinated by insects. Although the large showy flowers are only receptive to pollination from the time they open in the morning until early afternoon, this is also the time when bees are most active. Because of this fact, it is best to grow only one variety of okra per season or be prepared to isolate each variety by 500 ft. to ½ mile.

Since many seeds are born within an okra pod, only a few pods are needed for the average home gardener. To ensure purity, enclose individual flowers in large, lightweight blossom bags that can encompass the entire opened flower. Place the bag on the flower(s) the night before they open and leave them on for at least 32 hours to ensure they have pollinated themselves. Blossoms that are ready to open are often swollen and show light green striping. Do not reuse the bags on another flower for at least 24 hours to prevent pollen transfer and be sure to tag the fruits for seed collection.

GENERAL HARVEST

Picking a few immature pods for table use from plants being used for seed production is acceptable, but keep in mind that if you wait until the end of summer to save seeds you are essentially selecting for late fruit production. I suggest saving seeds early and reaping the edible harvest afterwards. Once mature seed-producing pods are removed, the plant should resume normal productivity well into the fall.

Okra pods are very large and inedible when ripe. Whenever possible, pods should be left to dry on the plant. To avoid shattering in the field and abundant self-sowing, collect the ripe pods just after they turn brown and place them in a paper bag until fully dry.

At this point, the fruits are very hard and may require a firm hand when flailing. Wrap pods in a heavy tarp and stomp on them with heavy boots or pound on them with a blunt instrument like a 2x4. The dust from okra pods can be very irritating to the skin, eyes and respiratory system, so wear gloves, long sleeves and a paper mask when threshing and winnowing.

OTHER

Okra is not related to, nor will it cross with Chinese okra (*Luffa acutangula*), also known as angled luffa, smooth luffa, sponge gourd, or vegetable sponge. Although okra is self-pollinating and viable seed can be had from just one pod, it also quickly succumbs to inbreeding depression, which often results in decreased germination and reduced production. For the home gardener, saving seed from 5–10 plants will avoid inbreeding depression. For preservation of a rare variety, expand that number to 25 plants or more.

Eggplant *(Solanum melongena)* will cross with all varieties of eggplant.

Hot and Sweet Peppers *(Capsicum annuum)* will cross with all varieties of pepper except Manzano peppers *(C. pubescens)*.

Garden Tomatoes *(Solanum lycopersicum) (formerly Lycopersicon esculentum)* will cross with all varieties of garden, cherry and currant tomatoes *(Solanum pimpinellifolium)*.

Tomatillo/Husk Tomato *(Physalis philadelphica)* will cross with all varieties of tomatillos and husk tomatoes.

Ground Cherry *(Physalis grisea) (formerly P. pruinosa)* will cross with all varieties of ground cherry, strawberry tomato, and dwarf cape gooseberry.

Currant Tomato *(Solanum pimpinellifolium)* will cross with all currant, garden and cherry tomatoes.

POLLINATION & ISOLATION

For all nightshades. Members of this family have perfect self-pollinating flowers that tend to pollinate themselves before opening. However, cross-pollination by insects is not only possible, but quite likely in most species. While peppers are cross-pollinated by insects more readily than modern tomato varieties, certain tomato varieties have flowers with a style that protrudes beyond the fused anther cone, making them available to insects. In general, tomato varieties should be separated by at least 10–20 feet, while very rare varieties and those with protruding styles should be separated by as much as 100 feet and a tall flowering crop between varieties. If space is limited, use wide rows of peppers as a barrier to tomatoes, as a barrier to eggplant, and so on. Where heavy pollinator activity is present, take extra precautions to prevent cross-pollination.

Eggplant *(Solanum melongena)*. Will cross with all varieties of eggplant.

GENERAL HARVEST

The fleshy fruits of eggplant are botanically mature when the skin begins to take on a brown or yellow hue and loses its shine. If possible, fruits should be left on the plants until they separate easily from the stem. If needed, fruits will continue to mature off the plant and can be kept in a warm place for an additional two to three weeks after picking.

Most of the minute, shiny black seeds of eggplant are imbedded in the flesh of the bottom third of the fruit. The hard, slippery seeds are not easily damaged, which helps with their removal. The easiest way to get the seeds out is to place large chunks of eggplant in a food processor with the dough blade attached. Add several cups of water and blend until pulpy. (If you don't have a food processor, you can reduce the flesh using a hand-held cheese grater.) Place the pulp into a sink or bucket and, using your hands, squeeze and mash until the seeds come loose. Once the majority of seeds are freed from the flesh, add more water and let the ripe seeds settle to the bottom. The pulp and unripe seeds will float and can be poured off. Repeat decanting as needed and spread the seeds on paper coffee filters to dry for 1–2 weeks.

OTHER

Eggplants are tender perennials grown as annuals. Viable seed can be obtained from just one plant, but to prevent inbreeding depression save seed from 5-10 different plants.

Hot and Sweet Peppers (Capsicum annuum). Will cross-pollinate with all pepper varieties except Manzano peppers (C. pubescens).

GENERAL HARVEST

Botanically ripe peppers are those that have achieved full coloration and have begun to soften. When possible, peppers should be allowed to fully ripen on the plant. If this is not possible, they should at least have a blush of ripe color before being picked and further ripened indoors. To harvest seeds, cut open the peppers carefully and strip the seeds from the white core and membranes. Rinsing is not necessary, but when the seeds are covered with water and gently stirred, mature seeds sink to the bottom while chaff and immature seeds float to the top. Pour off the chaff, repeating as necessary, and drain the seeds in a mesh strainer set on absorbent toweling. Spread the seeds out on a paper coffee filter to dry. Avoid drying in direct sunlight until the seeds break when folded in half.

OTHER

Until recently, botanists believed that the various species of *Capsicum*, which have perfect self-pollinating flowers, did not readily cross-pollinate. These days, we know that the most common species of peppers not only cross-pollinate rather easily, but also have a large array of mating systems that range from self-fertile to self-incompatible and that these systems can change based entirely on environmental conditions. In short, all pepper varieties should be treated more like out-crossers than inbreeders. The first four species in the following list make up what is known as the Capsicum annuum Complex and includes four distinct pepper species that can and often do cross-pollinate one another. The last pepper on the list, *C. pubescens,* is the only one that will not cross-pollinate with any other species but its own.

Common Sweet and Hot Peppers (*C. annuum*)

Tabasco and Squash Peppers (*C. frutescens*)

Chinese Lantern Peppers (*C. chinense*)

Kellu-Uchu Peppers (*C. baccatum*)

Manzano Peppers (*C. pubescens*)

In peppers, hot genes dominate those of sweet. In addition, genetic changes do not always manifest themselves in the first or even second generation of offspring. Because of this, seed producers often separate different varieties of sweet and hot peppers by as much as ¼ of a mile, but for the home gardener, isolating varieties by 20 ft. should be sufficient. At the very least, isolate sweet varieties from hot ones by planting them at opposite sides of the garden. Because peppers are often pollinated by bumblebees and sweat bees, care should be taken to strictly isolate rare varieties.

Tomatillo/Husk Tomato *(Physalis philadelphica)*. Will cross with all varieties of tomatillos and husk tomatoes, but not with ground cherries *(P. grisea)*.

GENERAL HARVEST

Tomatillos are fully ripe when the fruit swells and the outer papery husk begins to split open. Often the fruits turn from green to yellow when ripe. Remove the husks, rinse the fruits well to remove the greasy-sticky coating on the outside and gently pulverize them in water using a food processor with a dough blade. Otherwise, fruits may be chopped, placed in a bit of water and the pulp squeezed by hand to release the many small seeds. Add enough water to float the chaff and immature seeds and allow ripe seeds to settle. Pour off the chaff and repeat as necessary. Tomatillo seeds do not need to be fermented. Dry until brittle.

OTHER

Tomatillos and husk tomatoes are tender perennials generally grown as annuals and one of the few garden crops that do not exhibit inbreeding depression. Once considered to be two distinct species, tomatillos and husk tomatoes are now grouped together as *P. philadelphica*. Both are very closely related to ground cherries *(P. grisea)* but will only cross with other varieties of *P. philadelphica*. All varieties in this species have perfect self-pollinating flowers that are self-incompatible, which means that they rely entirely on insects for pollination. This also means an increase in isolation distances of 800 ft. or more between different varieties.

Tomato *(Solanum lycopersicum)*. Will cross with all varieties of common, cherry, and currant tomatoes *(S. pimpinellifolium)*, but not with tomatillos, husk tomatoes or ground cherries.

GENERAL HARVEST

Allow tomatoes to ripen fully before harvesting, either on the plant or in storage. Cut beefsteaks and paste tomatoes in half between the stem and blossom end. Squeeze the seeds, gel and juice into a large jar or glass bowl. Currant and cherry tomatoes can be pulverized in a food processor using a plastic dough blade. Do not add water to the fermentation vessel unless absolutely necessary, as seeds may sprout during fermentation. Ferment the seeds following the instruction in Chapter 5, Cleaning Wet Seeds and Fermentation. After cleaning, spread seeds thinly on a clean paper coffee filter out of direct sunlight and stir frequently to prevent clumping as they dry. Seeds that are completely dry will break cleanly when folded in half.

OTHER

Forget everything you have ever heard or read about tomatoes and cross-pollination. All tomatoes have a style (the female flower part) within the fused anther cone at the very center of the flower and the position of the style can differ radically. Some plants may have styles that are completely enclosed (inserted) in the anther cone, while others will have styles that completely protrude (exerted) beyond the cone. Others may have styles that are only partially exposed or are flush with the open-end of the anther cone. Only plants with fully inserted styles are perfectly self-pollinated. All others are prone to cross-pollination by insects. The type of tomato you grow, whether modern, heirloom, or potato leaf, does not determine the position of the style. In fact, the style can differ between similar varieties and can actually change during the growing season due to pollinator pressure and environmental conditions. Only close observation of the style will tell you whether a variety is prone to outcrossing or not. To be on the safe side, isolate all tomato varieties from one another by 10–20 ft. and up to 100 ft. for the preservation of rare varieties. Never save seed from plants with double flowered fruits, as this is a genetic flaw. Rogue out late germinating or weak plants.

challenging

The following crops can be saved by both beginners and more experienced seed savers. Cucumber, melons, muskmelon, spinach, squash and pumpkin are not self-pollinating and rely entirely on wind or insects for pollination. For these reasons, these crops require isolation techniques such as single variety plantings, staggered bloom time, greater isolation distances, hand-pollination, or caging to avoid cross-pollination. For the beginner, the easiest method of ensuring seed purity is to grow only one variety of each species in the garden at any given time. For example, grow only one variety of watermelon, muskmelon, cucumber and squash per garden, per season. These crops present the grower with an opportunity to work with another gardener to save multiple varieties of the same species.

CHENOPODIA

Chenopodiaceae Family

Spinach *(Spinacia oleracea)* will cross with all other varieties of common spinach.

POLLINATION & ISOLATION

Spinach plants are either all male or all female (dioecious). Male plants produce super fine pollen that can be carried on the wind for up to 5 miles. Common spinach may cross with wild varieties and those grown by close neighbors.

Because it is difficult to tell male plants from female plants until the flowers form, plant spinach in thick blocks and after flowering occurs, thin to one male for every two females. Spinach requires up to 6 weeks beyond eating stage to produce flowering stalks, therefore, it is best to start or plant spinach as early as possible in the season.

In mild winter areas, spinach may be planted in fall. Plants grown for seed should be spaced 8" apart as crowded plants bolt prematurely. Spinach begins bolting when day lengths reach 12–15 hours. Oftentimes, male spinach plants bolt before female plants. Take care when roguing out.

Once the flowers have formed, tie several female and male flowering stalks together and bag with heavy garden-grade paper, making sure to block the bottom of the bag with batting to prevent pollen escape and insect entry. Do not use spun polyester or other fabrics as bagging material because the pollen will move easily through the mesh. Tape or tie bags as necessary. Shake the bag daily to distribute pollen until flowering is complete.

GENERAL HARVEST

After flowering, allow the plant to dry in the field if possible, otherwise, pull the plants from the ground and hang upside down in a shady, dry area until completely dry. Wear gloves to strip the pods from the stems onto a tarp and flail, if necessary. Seeds should be completely dry before storing.

OTHER

The shape and texture of spinach seeds can help determine the type of leaves that will be produced. Smooth seeds produce plants with wrinkled leaves and prickly seeds produce plants with flat leaves. Harvesting a small quantity of the early leaves for the table should not hurt flower production.

Common spinach will not cross with orach mountain spinach (*Atriplex hortensis*), or unrelated plants with similar names, such as spinach mustard (*Brassica rappa*), Malabar spinach (*Basella alba*), water spinach (*Ipomoea aquatica*), or New Zealand spinach (*Tetragonia tetragonoides*).

CUCURBIT

Cucurbitaceae Family

Cucumber *(Cucumis sativus)* will cross with all varieties of common cucumbers.

Melon *(Cucumis melo)* will cross with all varieties of muskmelon, cantaloupe, honeydew, Armenian cucumber, pocket melon, vine peach, casaba and pickling melons.

Watermelon *(Citrullus vulgaris)* will cross with all varieties of watermelon and citron melon.

Squash and Gourds *(Cucurbita species)* will cross within each species.

Special Note on Cucurbits. The members of this large family are mostly monoecious; having separate male and female flowers on each plant. Most members of this family have bright, conspicuous flowers that are very attractive to pollinators and cross-pollination among species within the same genus should be expected. Female flowers are distinguished by a slightly swollen ovary beneath the petals, which resembles a miniature fruit. To keep seeds pure, grow only one variety of each species or be prepared to hand-pollinate.

Cucumber *(Cucumis sativus)*. Will cross with all varieties of common cucumbers.

POLLINATION & ISOLATION

Cucumber fruits grow best size allowed to freely sprawl on the ground or when trained to climb a trellis. Use mulch to avoid rot and leaf diseases. Expect male flowers to open as many as two weeks before female flowers, which are produced in abundance when day lengths reach 11 hours.

For pure seed, grow only one variety of cucumber per season or be prepared to either hand-pollinate or to isolate different varieties by up to ½ mile. Cucumbers are heat sensitive and will drop flowers and fruit in very hot weather. Hand-pollination should be done during the cool hours of the morning and early in the season. Use at least two male flowers for each female flower, preferably from different plants. Loosely tag the stems of pollinated flowers with colored string and if the plants are allowed to sprawl, place a tall stake beside each pollinated flower for easy relocation.

GENERAL HARVEST

Cucumbers reach botanical maturity approximately 5 weeks after market maturity and are large, soft and yellowish-red or brown in color. Fruits that are beginning to turn color can be removed from the plant and ripened on a counter indoors for an additional two weeks.

Cut the cucumber in half from the stem to blossom end and scrape out the seeds, gel and some of the watery flesh into a large glass jar. If needed, add a tiny amount of water to the jar and ferment for 3–4 days, stirring once or twice a day. Decant the seeds in clean water and pour off the immature seeds and chaff. Most of the ripe seeds will sink when immersed in water. Dry on paper coffee filters until seeds break cleanly when bent in half.

OTHER

To avoid inbreeding depression, save seeds from a minimum of 6 different cucumbers on 6 different plants. Better yet would be to save seed from 1 cucumber from up to 12 plants. It is possible harvest young cucumbers for the table without inhibiting those being grown for seed.

The cause of bitterness in cucumbers is two-fold. Some occurrences are environmental, caused by stress, heat and drought. This type of bitterness is weather related and dissipates when conditions change. However, there is also a genetic link to bitterness that can be perpetuated in seed saved from plants that produce bitter cucumbers. So, it is best not to save seed from any plant that consistently produces bitter fruit. Also, rogue out any plant that is misshapen, slow to germinate, diseased or unproductive.

Cucumbers belong to the very large and complex genus *Cucumis,* which includes two species that produce cucumbers, melons and melon-like fruits such as Armenian cucumbers and snake melons. But don't let names, or looks, fool you; only *Cucumis sativus* produces traditional slicing and pickling cucumber varieties and they will only cross with other cucumbers belonging to the species *sativus.* They will not cross with Armenian cucumbers and snake melons (*Cucumis melo*) or true gherkins (*Cucumis anguria*). Snake gourds, also known as serpent or club gourds (*Trichosanthes anguina*) belong to an entirely different genus and will not cross with cucumbers.

Melons (*Cucumis melo*). Will cross with all varieties of cantaloupe, muskmelon, honeydew, Crenshaw, casaba, vine peach, Armenian cucumber and melon apple, as well as Persian-, pocket-, Asian pickling-, orange-, mango- and lemon melons, among others.

POLLINATION & ISOLATION

Inbreeding depression is rare in melons, but all *melo* species will readily cross-pollinate one another. They will not cross with watermelons (*Citrullus vulgaris*), which belong to a different genus altogether. To prevent cross-pollination within the species, grow only one variety at a time, or separate different varieties by ½ mile. Otherwise, hand-pollinate individual flowers.

Melons tend to abort roughly 80% of all female flowers and only about 10% of hand-pollinated flowers will result in mature fruits. The most receptive flowers are those that open early in the season and it is advisable to hand-pollinate early and often. Use at least two male flowers to pollinate one female flower. Loosely tag the stems of pollinated flowers with colored string and place a tall stake beside them for easy relocation.

GENERAL HARVEST

Harvest melons for seed when at the very ripe, edible stage. If you are unsure whether a melon is ripe, allow it to mature off the vine until the rind begins to soften and flex to the touch. Cut the melon in half between the stem and blossom end and scrape the seeds into a large jar. Fill the jar with water, cap and shake vigorously until the seeds come free from their placental membranes. Allow mature seeds to settle to the bottom, pour off floating chaff and unripe seeds and repeating as necessary until clean. Strain and dry on paper coffee filters away from direct sunlight until seeds break when folded in half.

Watermelon *(Citrullus vulgaris)*. Will cross with all varieties of watermelon and citron melon.

POLLINATION & ISOLATION

Watermelons have male and female flowers on each plant and will cross with all watermelons and citron melons, but not *C. melo* varieties such as cantaloupe or honeydew melons. To ensure pure seed it is best to grow only one variety of watermelon or citron at any given time. Otherwise, isolate varieties by ½ mile or hand-pollinate individual flowers. Early flowers are the most receptive to hand-pollination and 75% of those that are hand-pollinated will produce fruit. The most receptive flowers are those that open after the initial flush of male blooms early in the season. Use at least two male flowers to pollinate one female flower. Loosely tag the stems of pollinated flowers with colored string and place a tall stake beside them for easy relocation.

GENERAL HARVEST

Harvest watermelons for seed when at peak market maturity. Various methods are used to determine ripeness: the yellow ground spot may soften in color, the tendril closest to the stem may dry up, the stripes on the rind may turn from solid to broken, and when thumped, the melon might sound hollow. The rinds of excessively ripe melons often lose all sheen and will flex with firm pressure.

For seed harvesting, select very ripe edible melons. Don't leave the fruit to ripen too long, or the seeds may actually sprout inside the fruit. The best method to determine ripeness is timing. When tagging stems for hand-pollination or after you've found a baby fruit just getting started, count forward the number of days to maturity and write the date on a tag with permanent marker and affix it to the fruit's stem. After harvest, allow an additional two weeks to ripen indoors. Scrape out seeds and rinse. Most of the ripe seeds will sink in water. Dry on paper coffee filters until brittle.

Squash and Gourds *(Cucurbita species)*. Will cross within the species as indicated below.

POLLINATION & ISOLATION

All squash species have separate male and female flowers on each plant and rely on insects to pollinate their very large and showy flowers.

There are conflicting reports of crossbreeding within the different species of the genera, but for the home seed saver, the best method of ensuring purity is to grow only one variety of each of the four species during the growing season. Otherwise, isolate varieties within the same species by up to a mile or hand-pollinate individual flowers.

Squash tend to produce only male flowers during the first flush of blooms, wait until a decent percentage of female flowers appear before hand-pollinating. Use at least two male flowers for each female flower. Use the spent male flower to cover the re-taped female flower, which adds another layer of protection against bumblebees, known to chew through unopened blossoms. Loosely tag the stems of pollinated flowers with colored string and place a tall stake beside them for easy relocation.

The following four species of *Cucurbita* generally do not cross with one another. To avoid cross-pollination within each species without hand-pollinating, grow only one *C. maxima*, one *C. argyrosperma*, one *C. moschata* and one *C. pepo* during the growing season.

Cucurbita maxima varieties include banana, buttercup, Hubbard, Hokkaido, kubocha, sweet keeper, red kuri, delicious, French turban, and marrows.

Cucurbita argyrosperma (formerly *C. mixta*) include all cushaws, many green-and-white striped squash, Japanese pie, silverseed gourd and Tennessee sweet potato.

Cucurbita moschata varieties include, cheese type squashes and pumpkins, all butternuts, and winter crooknecks.

Cucurbita pepo varieties include many types of gourds, winter squashes such as acorn, delicata, cocozelle, English marrow, most types of sweet pumpkins, and all summer-type squashes, such as yellow, scallop, spaghetti and zucchini.

GENERAL HARVEST

For all squash, including summer types, the fruits are mature when the outer shells have hardened and a fingernail pushed into the rind does not leave a dent. Cut the fruit from the vine and further ripen for 3-4 weeks. In the case of long-keeping winter squash, seeds can be harvested when the flesh is prepared for the table.

For gourds, it is best to harvest seeds before the outer shell becomes brittle. Cut the fruit lengthwise, from stem to blossom end and scoop out the seeds into a large bowl or bucket. Add a generous amount of water and separate the seeds from the fleshy membranes. Rinse seeds thoroughly in clean water, pour off unripe seeds and drain in a mesh strainer. Dry on paper coffee filters away from direct sunlight until the seed breaks when folded in half.

OTHER

Select squash for vigor, disease resistance, fruit color, shape and size. Do not save seed from weak, diseased or slow germinating plants.

conclusion

i have been gardening and saving seed for over a quarter of a century. What started with a tomato seed here or way to save a buck there went from quirky curiosity to a seed saving passion with deep and abiding roots. After all these years, it's truly exciting to see so many people interested in saving their own garden and market seed. To those who believe in seed sovereignty, eco-agriculture, and food freedom, a seed saving fetish is exactly what this world needs right now.

It's worth revisiting, then, Thomas Jefferson's own passion for cultivating diversity when he said, "The greatest service which can be rendered any country is to add a useful plant to its culture." A lover of botanicals and cultivating wholesome foodstuffs, Jefferson surely did his part to add and preserve as many useful seeds to our foundling country as he did plants. It's up to us now to revitalize and preserve our natural legacy of seed, as well as our right to grow and save them without penalty or repercussion.

I hope that through this book I can help my fellow gardeners lean how to save their own high-quality open-pollinated seed at home. And while the details might seem a bit overwhelming at first, the truth is that anyone can save seed and do it easily and well. Once you begin to save your own seed, I dare say you will never stop.

Happy seed saving and may all your seeds come true!

sources

Allard, R.W. *Principles of Plant Breeding*. John Wiley, Inc., 2nd Ed, 1999.

Anderson, Luke. *Genetic Engineering, Food & Our Environment*. Chelsea Green, 1999.

Ashworth, Suzanne. *Seed to Seed: Seed Saving Techniques for the Vegetable Gardener*. Decorah, Iowa. Seed Savers Publications, 1991.

Bubel, Nancy. *The New Seed-Starters Handbook* . Rodale Press, 1988.

Deppe, Carol. *How to Breed Your Own Vegetable Varieties: The Gardener's and Farmer's Guide to Plant Breeding and Seed Saving*. Chelsea Green Publishing, 1993.

Facciola, Stephen. *Cornucopia: A Sourcebook of Edible Plants*.

Klopenburg, Jack. *Seeds & Sovereignty*.

McCormack, Jeff, Ph.D. Article: *Guidelines for Maintaining Purity in Pepper Varieties*. Southern Exposure Seed Exchange. www.southern exposure.com

Rural Advancement Foundation International, USA. (RAFI-USA)

Whealy, Kent. *The Garden Seed Inventory*. Seed Savers Exchange.

ill Henderson is an artist, author, self-taught herbalist and naturalist. Her first book, entitled *The Healing Power of Kitchen Herbs*, focuses on the simplicity of growing and using the world's most common herbs for both food and medicine.

A life-long organic gardener and seed saver with a passion for sustainable agriculture and local food production, Jill teaches workshops that teach experienced and beginner gardeners about the current global challenges presented by bio-engineered food crops and how participants can grow and save their own open-pollinated and heirloom seeds.

Jill and her husband, Dean, live in the Missouri Ozarks where they produce organic herbs, small fruits and vegetables on their rural homestead. Their farming practices focus on growing regionally adapted varieties with an emphasis on natural soil enrichment, sheet composting, green manures and deep-mulch techniques. Pest and disease control include crop rotation, companion planting, diversion crops and the use of dusts or sprays made with organic herbs.

In 2010, Jill launched Show Me Oz, (ShowMeOz.wordpress.com) a weekly blog featuring articles on nature, conservation, sustainability, gardening, seed saving, homesteading, alternative health and more.

In her spare time, Jill works as an artist specializing in custom pet portraits and wildlife art. You can view samples of her work at ForeverPetPortraits.wordpress.com.

Connect with Jill on Facebook
Simply type "Show Me Oz" in the search

SHOW ME
OZ
.WORDPRESS.COM

other works BY JILL HENDERSON

THE HEALING POWER OF KITCHEN HERBS

Growing & Using Nature's Remedies

Be prepared for the changing times with this no-nonsense guide to growing, propagating, and using 35 of the world's safest, most effective, and flavorful kitchen herbs. Learn how to season food and create useful and safe herbal remedies for everyday use. Written by a veteran gardener and home herbalist, *The Healing Power of Kitchen Herbs* is a treasured resource that you will turn to again and again.

Available in paperback at createspace.com/3477724
or in e-book formats at https://www.amazon.com/Jill-Henderson/e/B005IDO2TE

A JOURNEY OF SEASONS

A Year in the Ozarks High Country

Take a walk on the wild side into the rugged heart of the Ozark Mountains with noted author and naturalist, Jill Henderson, as she entices you to look deeper into the mysterious and endless circle of life. Filled with nature notes, botanical musings, backwoods wisdom, and just a pinch of hillbilly humor, this is one journey you'll never forget.

Available in paperback at createspace.com/3477718
or in e-book formats at https://www.amazon.com/Jill-Henderson/e/B005IDO2TE

GROUNDSWELL BOOKS
SOLUTIONS FOR A SUSTAINABLE WORLD

For more books that inspire readers to create a healthy,
sustainable planet for future generations, visit
BookPubCo.com

How to Start a Worm Bin
Henry Owen
978-1-57067-349-8
$9.95

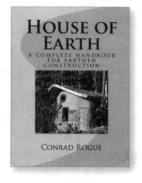

House of Earth
Conrad Rogue
978-1-53064-281-6
$14.95

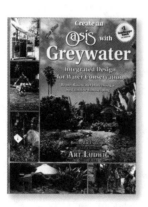

**The New Create an Oasis
with Greywater**
Art Ludwig
978-0-96434-333-7
$22.95

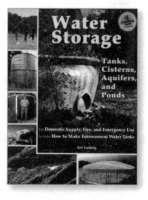

Water Storage
Art Ludwig
978-0-96434-336-8
$19.95

Purchase these titles from your favorite book source or buy them directly from:
Book Publishing Company • PO Box 99 • Summertown, TN 38483 • 1-888-260-8458
Free shipping and handling on all orders.